MW01343945

WHAT IS VOLCANOLOGY?

Volcanology is the study of volcanoes, including how they are formed, why they erupt, and how to predict future eruptions.

The scientists who study volcanoes are called **VOLCANOLOGISTS.**

CAN VOLCANOES TURN THE MOON BLUE?

DISCOVER THE SCIENCE BEHIND ***VOLCANOLOGY***
(vol-cuh-NOH-luh-jee)

Written by Olivia Watson
Illustrated by Verónika Cháves Morales

Words that are tricky to understand are in **bold**. Find out what they mean in the glossary.

Words that are difficult to say are in *italics*. Find out how to say them at the back of the book.

More than 100 years ago, on a small island off the coast of *Indonesia*, a huge bang echoed and, in the distance, a cloud of smoke rose up, filling the sky. What had happened, and why was the Moon turning blue?

A huge volcano called *Krakatoa* had erupted.

It was devastating!

Lava, rocks, and **ash** came rushing out of the volcano, and the ash was spread by the wind. The eruption was so strong that it caused a **tsunami**! Sadly, many people lost their lives. Today, *volcanologists* say this was one of the most destructive eruptions ever.

This wasn't the first time the Moon had turned blue, and it wouldn't be the last time either!

This unusual event had been happening for thousands of years, often after a volcanic eruption. But could it really be volcanoes causing this?

Volcanologists today know that volcanoes have a lot going on beneath the surface. These mighty mountains stretch deep underground, where **molten** rock called **magma** bubbles in **chambers**, terrifyingly hot.

Magma and **gas** moving around inside the volcano can sometimes make noise. This can often tell volcanologists what is going on inside.

Aboveground during an eruption, volcanoes can shoot out rocks the size of cars, as well as burning hot lava that moves at incredible speeds. Some volcanoes even create lightning! They are

powerful forces of nature!

Like many things in nature, not all volcanoes are the same. Volcanologists have discovered more than 1,500 volcanoes on Earth, and most of them are deep down at the bottom of the ocean.

When underwater volcanoes erupt, lava can burst out of the ocean! As it touches the water, lava cools and turns into solid rock. Powerful eruptions can create new islands, completely changing Earth's surface!

Volcanoes are full of surprises – they could erupt at any time! Some stay **dormant** for thousands of years before erupting again.

That's why volcanologists are so important. These brave scientists work hard to understand volcanoes so they can learn to predict eruptions sooner and help keep people safe. So, what actually causes a volcano to erupt?

Eruptions happen because of a buildup of gas and **pressure** inside the magma chamber. When it's too much for the ground to hold, it erupts!

Working with other scientists, volcanologists found out that this pressure creates small **earthquakes**. By using clever tools that can show **tremors**, scientists can detect signs that an eruption might happen soon.

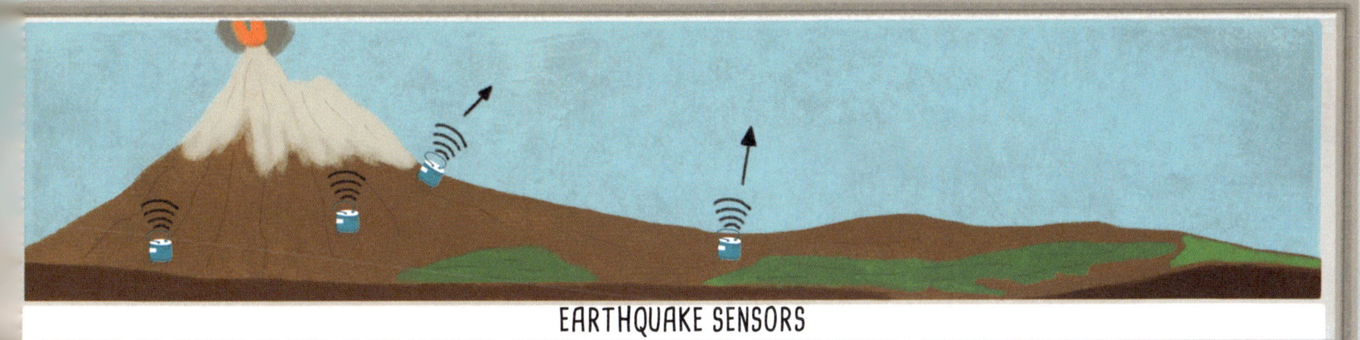
EARTHQUAKE SENSORS

INSIDE THE CRATER

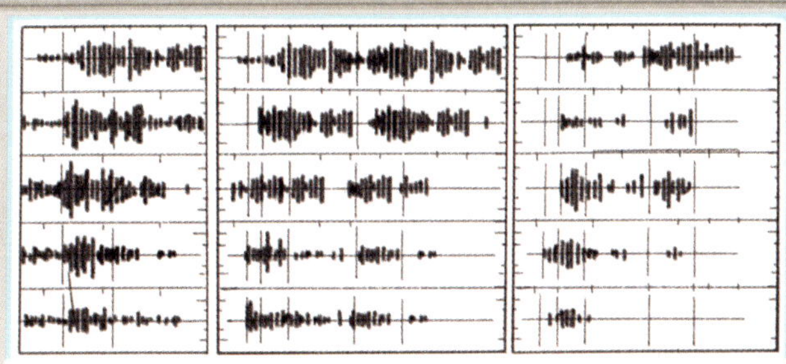
RECORDED TREMORS

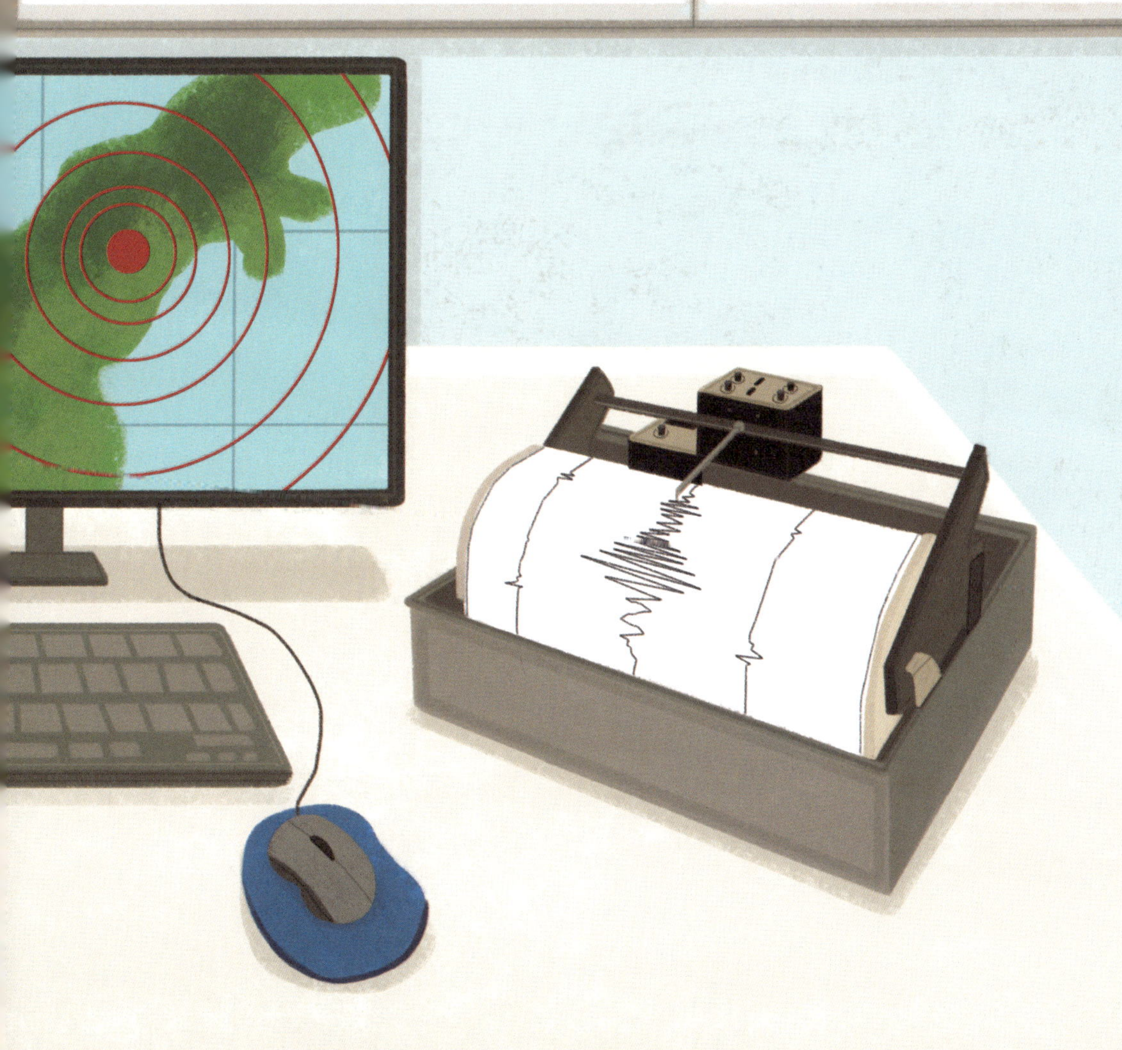

But earthquakes aren't the only signs that a volcano might be preparing to blow! Volcanologists monitor any changes to the ground or air around the volcano too.

They use **drones** to test the gas coming out of the **crater** and special cameras to monitor underground heat. They even watch for new lumps on the side of the volcano, which could be magma building near the surface!

Volcanoes are powerful, but they're not always bad! In fact, by learning how to predict eruptions, we can make use of the positive things they do.

Volcanic ash improves soil which helps crops grow...

volcanic rock holds **precious** materials like diamonds and gold, and can also be used to make tools, **electronics**, and cleaning products...

and scientists even think we can take the power from volcanoes and turn it into **clean energy**.

Volcanologists today have a system for showing the strength of eruptions. They were able to work out that Krakatoa's big eruption was a 6 out of 8.

It was so strong that people in other countries heard it erupt! And the force of the eruption sent a cloud of ash really high into the sky...

7 out of 8
Mt Tambora, 1815
8 out of 8
Yellowstone, 640,000 BCE

This ash was picked up by the wind and spread around the world.

From studying volcanic ash, scientists today know that ash **particles** react with the light coming off the Moon, causing it to look blue.

While it's true that volcanoes can make the Moon look blue, it doesn't stay this way forever, it's just how we see it for a little while. Because the Moon is so far away there's no way the ash could reach it. But it's still

amazing to see!

Extreme

VOLCANOES

Now we know all about Krakatoa, what other volcanoes are out there? Here's just a few of the most powerful ones!

MOUNT VESUVIUS, *Italy*

When Vesuvius erupted in 79 CE, the city of Pompeii was covered in lava, ash, and mud so thick that it preserved buildings, items, and even people's bodies!

YELLOWSTONE, *USA*

Located in a national park, "supervolcano" Yellowstone covers a huge area of land that is very active, often experiencing **geysers** shooting hot steam high into the air. The volcano has had 3 gigantic eruptions in the past!

ANAK KRAKATOA, *Indonesia*

Since Krakatoa's destruction in 1883, a new volcano has risen from the sea in the same place. Indonesians call this volcano "Anak Krakatoa", which means "Child of Krakatoa".

PACIFIC RING OF FIRE, *Pacific Ocean*

A lot of the world's volcanoes and almost all of its earthquakes are in the Pacific Ring of Fire, a horseshoe-shaped area of the Pacific Ocean. Krakatoa can be found here.

EYJAFJALLAJÖKULL, *Iceland*

A huge ash cloud from the eruption of *Eyjafjallajökull* in 2010 – an eruption which lasted 71 days – caused thousands of planes to be stranded for several weeks.

Explosive

VOLCANO FACTS

Volcanoes are powerful things that can make the Moon look blue, but what else is there to know about these mighty, fiery mountains?

KRAKATOA HAS ANOTHER NAME!

The name "Krakatoa" was famously written down wrong! While this spelling is used in the English-speaking world, in Indonesia, it is written as "Krakatau".

VOLCANOES CAN MAKE THE WORLD COLDER.

In the past, huge volcanic eruptions have pushed ash so high that it covered the sky like a thick blanket, blocking out the Sun's warmth and making the world colder!

VOLCANOES ARE THE LOUDEST THING ON EARTH!

The 1883 eruption of Krakatoa (the main one in this book) was the loudest sound in recorded history. It was so loud that it was heard in really faraway countries.

EARTH ISN'T THE ONLY PLACE WITH VOLCANOES!

The Moon and Mars are covered in volcanoes (although they're no longer **active**). In fact, the biggest volcano in the solar system is on Mars. It's called "Olympus Mons".

ANIMALS CAN TELL WHEN AN ERUPTION IS COMING!

Some animals, like dogs and goats, start acting strangely before an eruption. Scientists think these animals might be able to sense ground movements or chemical changes that humans can't.

GLOSSARY

Active (volcano) – erupting or likely to erupt.

Ash – a powder that is left behind after something is burned.

Chambers – large pockets of space below the Earth's surface that collect magma (see right).

Clean energy – energy that does not harm the planet.

Crater – a hollowed-out area. Volcano craters are at the top of the mountain and are where lava comes out of.

Dormant – something that can go for a long time without any activity or growth.

Drones – small aircraft that don't carry people and usually have cameras attached. They are guided by remote control.

Earthquakes – when parts of the ground shake. They can be very strong and destructive.

Electronics – devices that use electricity to do lots of different things, such as phones and TVs.

Gas – tiny, usually invisible, particles in the air.

Geysers – a spring that throws up jets of hot water and steam from the ground.

Lava – hot, melted rock that has come from underground. It is called magma (see below) when it is still underground.

Magma – hot, melted rock that is underground. It is called lava (see above) once it is aboveground.

Molten – another word for melted.

Particles – tiny pieces of something.

Precious – something that is special, rare, or that people care for.

Pressure – a strong force that acts on something.

Tremors – a small shaking sensation. Tremors of the ground usually happen before an earthquake (see left).

Tsunami – a big sea wave, usually caused by an earthquake or volcanic eruption. *Need help saying this? Look below!*

HOW DO I SAY?

Eyjafjallajökull
AY-yaf-yat-lah-YOH-kull

Indonesia
IN-du-NEE-zee-uh

Krakatoa
crack-uh-TOE-uh

Tsunami
soo-NAH-mee

Volcanologist
vol-cuh-NOH-luh-jist

Volcanology
vol-cuh-NOH-luh-jee

THE BIG QUESTIONS ANSWERED

**This is more than just a series of books; it is a complete resource.
Accompanying each book is a variety of FREE material to engage curious kids with science.**

www.thebigquestionsanswered.com

Use the QR code to visit the website, download free resources, and discover other books in the series.

On the website, find out incredible things about volcanologists, including what they do, some of their greatest discoveries, and the people who have made a difference in this field of science.

The material is also available for home or classroom use, supporting all the information in this book.

Teachers' & Parents' Resources
With discussion prompts and questions, extra information, and facts around key topics.

Young Volcanologists' Activity Pack
Fun activities for wannabe volcano experts, including creative writing, drawing, word searches, and much, much more.

**The Big Questions Answered is published by Beetle Books.
Beetle Books is an imprint of Hungry Tomato Ltd.**

First published in 2024 by Hungry Tomato Ltd
F15, Old Bakery Studios, Blewetts Wharf, Malpas Road,
Truro, Cornwall, TR1 1QH, UK.

ISBN 9781835691359

A CIP catalog record for this book is available from the British Library.

With thanks to:
Editor: Holly Thornton
Editor: Millie Burdett
Senior Designer: Amy Harvey
The team at Beehive Illustration

Printed and bound in China.

Picture Credits:
(t = top, b = bottom, m = middle, l = left, r = right)
Shutterstock: Artsiom P 35mr; CGS Graphics 35tl; oleJohny 34bl; satoriphoto 33ml; Thorsteinn Asgeirsson 33br; Tunatura 32mr; RethaAretha 34mr; Wirestock Creators 35bl.